Solomon's Mystery

1 Kings 3:16

Coloring
PAGE

Solomon's Mystery
1 Kings 3:16

Solomon was promised wisdom, riches, honor, and long life if he would continue in righteousness before the Lord. The promise was fulfilled. During his life, Solomon became famous for his wisdom. ... Solomon also acquired great wealth, and there were said to be no kings in all the earth who could compare to him.

CSI
Solomon's Mystery

Atomic Kidz©™ STEM Series

Urban Missions Network, Inc
Urban Missions Institute
P.O. Box 250718
Atlanta, GA 30325
is a 501 (c) 3 organization

This book belongs to

_____

Solomon's Mystery

1 Kings 3:16

Two women came to Solomon and asked his judgment over a dispute. The two women both recently had young babies. The baby that belonged to one woman died so she switched her dead child with the living one of the other woman. She now claims that child is her own. The other woman said the child is hers and the other woman said the story was total nonsense.

Solomon then said the baby should be cut in half with a sword and each half given to one woman. The first woman said for him not to do this and to let the child live. She was willing to give up her claim of the child to let it live. The other women said to let it be divided.

Solomon gave the child to the woman who said to let it live. Only the true mother would say and do such a loving act to spare her child's life.

The Quest to Solve Solomon's Challenge

(1 Kings 3:25)

What problem did King Solomon face when two women claimed to be the mother of the same baby?

How did King Solomon's solution reveal the true feelings of the women?

Why do you think the real mother reacted the way she did to Solomon's plan?

What does this story teach us about wisdom and justice?

How might King Solomon's decision be applied to solve disputes in modern times?

OBSERVATION CHALLENGE:

Can you Spot the Difference?.

SCIENTIFIC METHOD WORD SEARCH

Look for the words listed below.

```
N  L  F  D  L  H  B  M  K  C  R  G  T  R  Y
E  O  I  D  P  T  X  E  P  E  Y  V  J  E  S
Q  H  I  S  A  Z  C  L  G  P  C  X  P  I  U
U  Y  R  T  T  S  W  B  R  L  V  E  S  N  M
I  A  X  U  S  I  N  O  I  M  F  E  A  L  M
P  C  T  X  Y  E  C  R  Q  C  H  B  H  F  A
M  F  O  A  I  E  U  P  S  T  L  U  S  E  R
E  K  Z  N  D  U  B  Q  O  F  Q  T  F  W  Y
N  O  T  U  C  I  N  P  W  Z  G  G  E  T  E
T  E  R  O  G  L  Y  M  A  T  E  R  I  A  L
R  E  M  B  D  H  U  Q  C  X  R  W  T  Z  B
S  H  X  U  Q  X  K  S  L  Q  K  B  N  M  A
E  A  H  J  B  T  K  E  I  H  R  Q  L  Y  T
A  N  A  L  Y  S  I  S  J  O  P  N  N  Q  D
H  Q  V  L  O  D  E  W  Z  J  N  C  S  P  G
```

Analysis	Conclusion	Data
Equipment	Hypothesis	List
Material	Problem	Procedures
Question	Results	Summary
Table		

CSI: THE MISSING I PHONE

Write what you feel, smell, see, hear and during your investigation.

QUESTION PROBLEM:

HYPOTHESIS:

SEE

SMELL

TASTE

FORENSIC EVIDENCE HUNT

Forensic Evidence Hunt I Spy Game:
Search for hidden clues related to forensic science within a
designated area, encouraging observation and deduction skills.
Connect each finding to the application of scientific principles.

"I spy with my detective eye, something that begins with..."

1. F: Find something related to fingerprints that helps identify
individuals.

2. M: Spot an item used by investigators to magnify and examine
tiny details.

3. C: Look for a clue marked with crime scene tape that signals
restricted access.

4. D: Discover an object used to collect and preserve evidence, often
made of clear material.

5. B: Identify an emblem worn by detectives that signifies authority
in investigations.

6. F: Locate a trace left at a crime scene, such as a shoe or footprint.

7. D: Spy an item used to analyze DNA, promoting advanced
forensic techniques.

8. M: Uncover a miniature model representing the structure of
genetic material.
ween biblical wisdom and scientific understanding.

Can You

SPOT THE DIFFERENCE?

Forensic Science Techniques

```
U R K R A P M I Q A J I D F H
S H H S N H A X U V X Y W N V
E C V S A L H T Q R G J Z A A
L P I E L I O C H O E X B R C
P V C T Y P J G L O I Y E F K
Y Z M N S W F O I S L G P T R
G C D Y I I T X H I U O U K V
U G S I S N L S A W F L G F U
D E N T O M O L O G Y O F Y D
L X C D T Z L O A R X R I L W
S F O P R H B P W B J E B P M
P Y G O L O C I X O T S E F U
U F R U N Z T J H G A J R A I
V P U Z V J K Q U Z N T S S P
M O Y X M V W Q E J Q E Z L R
```

DNA	SEROLOGY
ANALYSIS	ODONTOLOGY
AUTOPSY	ENTOMOLOGY
TOXICOLOGY	PATHOLOGY
BALLISTICS	FIBERS

CSI MEMORY CHALLENGE

Question: What were the features of the Crime scene?

Draw the other half of the evidence:

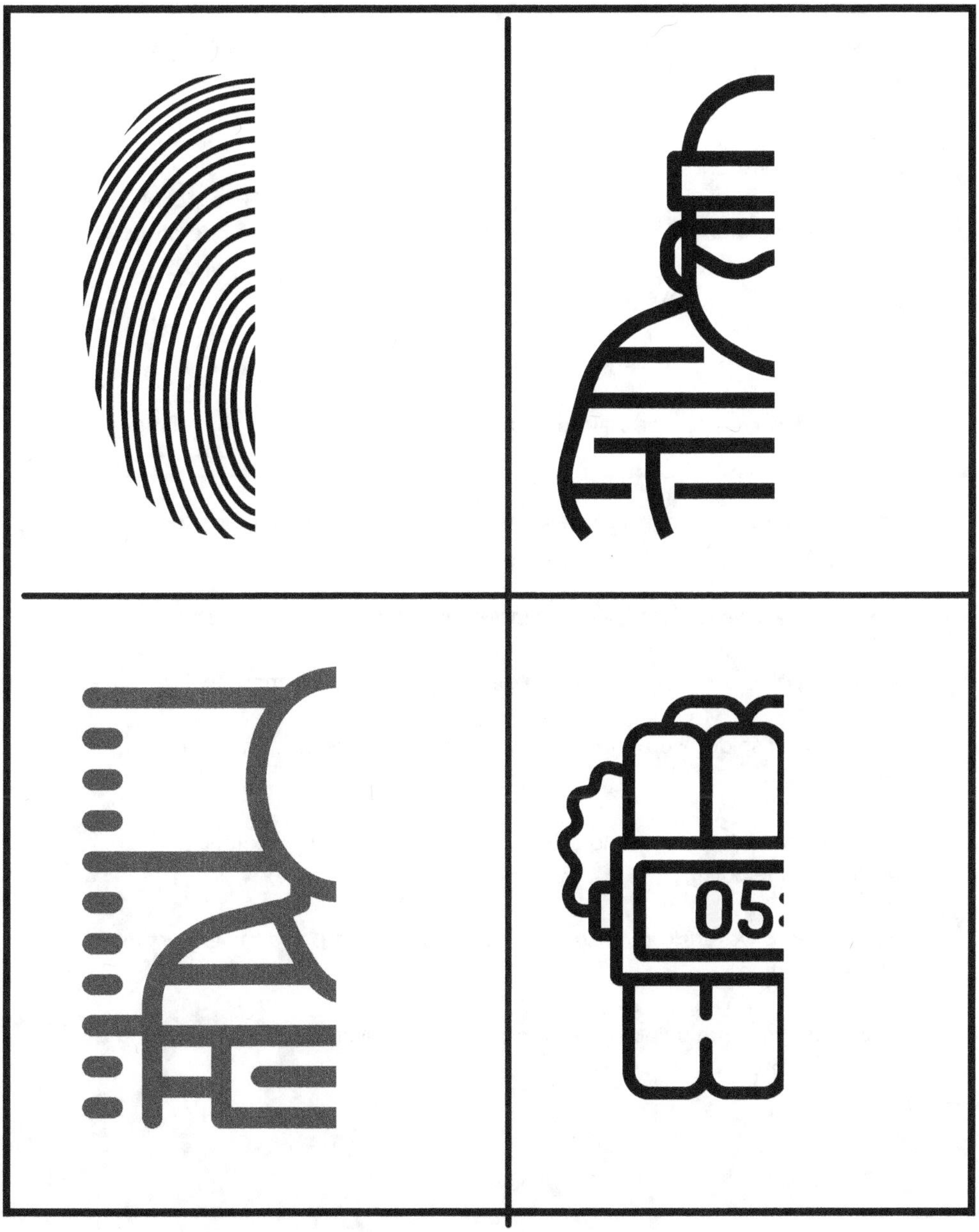

WISDOM IN NATURE SCAVENGER HUNT

Explore the outdoors and identify elements of nature that symbolize wisdom from King Solomon's story. Relate each find to a STEM concept, fostering a connection between biblical wisdom and scientific understanding

Wisdom in Nature Scavenger Hunt I Spy Game:

"I spy with my wise eye, something that represents..."

O: Find an owl figurine or an image of an owl, symbolizing wisdom in many cultures.

T: Locate a tree with rings, representing age and the wisdom that comes with time.

B: Spot a beehive, symbolizing organization and order within a community.

A: Identify an ant trail, representing diligence and cooperation.

S: Discover a spider web, symbolizing intricacy and the design found in nature.

R: Find a rock with interesting patterns, connecting to the geological wisdom of the Earth.

W: Uncover a water droplet, symbolizing purity, clarity, and the importance of conservation.

P: Seek a plant with unique growth patterns, representing adaptation and growth in nature.

SKILLS & TOOLS OF AN INVESTIGATOR

```
D U F H K I S P X A J U D P X
Y Q L L A Y E Z M O N H B R O
N Q F E Z N S L T L N G V J J
G H N N C J N W U C N U L Z U
N N T I O E E Q U I P M E N T
I O L O L I S Q Y N E P S E E
R T R P O N T R Z G A T F J S
U E O R H L O A D T R B V Y T
S B T I E M S R V I D O O L B
A O C K E L X K P R R L C G U
E O E M Y G U S L C E M U D J
M K T J Q N W R Z B T S W G R
M A E I S V Q Y J A L P B R U
Z L D Y P U F N R A O J F O N
M A G N I F I E R Y C N T W S
```

Blood	Pencil
detector	pH
Equipment	Ruler
Magnifier	Senses
Measuring	strips
Memory	tape
Notebook	test
Observation	Tools
Pen	

CATCH THE COMPUTER THIEF!!
FIND THE CYBER THIEF

Observation Challenge: What do you see?

The challenge: See how many missing items can you find?
Hint: There are (4) items.

The challenge: See how many
missing items can you find?
Hint: There are ( ) items

CSI Matching Challenge

Draw a line to the matching definition

Autopsy	Unwanted pollution of a crime scene, which can obscure evidence
Ballistics	The study of disease and its effects on the body
Contamination	Genetic blueprint used for identification in forensics
DNA	The use of insects to help solve criminal cases
Entomology	Postmortem examination to discover the cause of death
Fibers	Unique pattern used for personal identification
Fingerprint	Study of teeth structure and dental work for identifying human remains
Odontology	The study of the effects of being fired on a bullet, cartridge, or gun
Pathology	Threads or filaments from fabric used as evidence

Memory Challenge - The Crime Scene

What do you remember?

Raising their hand

Making fun of a classmate

Littering in the classroom

Running in the classroom

Sharing with a classmate

Helping a classmate

Saying thank you

Saying please

Not listening during class

CSI HUNT

```
E  W  T  R  F  C  I  S  N  E  R  O  F  D  S
C  C  E  B  G  S  U  Y  B  E  G  E  N  N  Y
N  R  E  G  H  C  B  J  H  T  E  A  R  R  S
E  I  H  J  E  G  Q  P  G  V  L  E  I  E  N
D  M  S  I  N  V  I  S  I  B  L  E  I  N  K
I  E  E  G  L  C  E  T  T  S  H  V  M  X  F
V  S  D  C  A  K  C  Z  W  Y  A  W  D  F  O
E  C  O  Z  N  E  B  G  V  M  R  M  V  P  O
Q  E  C  R  T  E  Q  B  I  B  L  C  U  I  T
L  N  M  E  B  B  D  T  M  O  H  Z  T  R  P
K  E  D  E  G  O  L  I  X  L  Z  R  N  D  R
H  E  J  I  Y  X  I  H  V  L  E  H  U  X  I
O  J  Y  P  X  Y  Z  F  E  E  O  R  H  E  N
Y  T  S  F  I  N  G  E  R  P  R  I  N  T  T
H  V  R  E  P  O  C  S  O  R  C  I  M  W  Y
```

Codesheet	Hunt
Cipher	Spy
Invisibleink	Fingerprint
Key	Crimescene
Symbol	Microscope
Puzzle	Evidence
Forensic	Detective
Evidence	Footprint
	DNA

draw:

REPORT

is my favorite season because...

_____

_____

_____

_____

_____

_____

_____

_____

_____

_____

FINGERPRINT IDENTIFICATION

Fingerprint Identification is the method of identification using the impressions made by the minute ridge formations or patterns found on the fingertips. No two persons have exactly the same arrangement of ridge patterns, and the patterns of any one individual remain unchanged throughout life. Fingerprints offer an infallible means of personal identification. Other personal characteristics may change, but fingerprints do not.

Fingerprints can be recorded on a standard fingerprint card or can be recorded digitally and transmitted electronically to the FBI for comparison. By comparing fingerprints at the scene of a crime with the fingerprint record of suspected persons, officials can establish absolute proof of the presence or identity of a person.

FINGERPRINT PATTERN TYPES

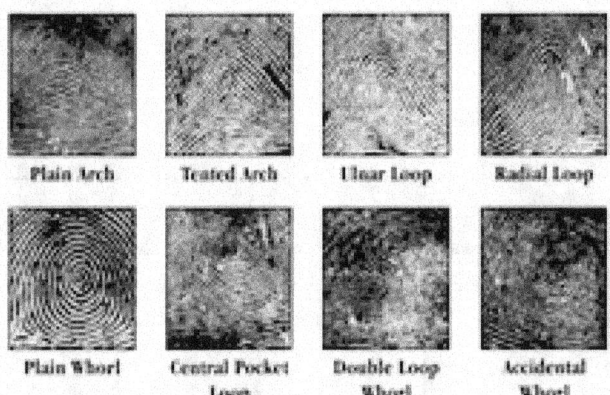

| Plain Arch | Tented Arch | Ulnar Loop | Radial Loop |

| Plain Whorl | Central Pocket Loop | Double Loop Whorl | Accidental Whorl |

EACH DAY APPROXIMATELY 7, 000 NEW INDIVIDUAL RECORDS ARE ADDED TO THE FILES.

Fingerprints are now processed through the **Integrated Automated Fingerprint Identification System.** The fingerprints are submitted electronically or by mail, processed on IAFIS, and a response is returned to the contributing agency within two hours or less for electronic criminal fingerprint submissions and twenty-four hours or less for electronic civil fingerprint submissions. Fingerprint processing has been reduced from weeks and months to hours and minutes with IAFIS.

FINGERPRINTING ACTIVITY
FOR BEGINNERS

MY RIGHT HAND

Thumb	Index	Middle	Ring	Pinky

MY LEFT HAND

Thumb	Index	Middle	Ring	Pinky

Do you have? Arches _____Loops _____ Whorls_____

Amazing DNA

THE SCIENTIFIC METHOD

Title: Make your title unique and memorable!

QUESTION PROBLEM:

The question/problem might be assigned to you, or you may have to come up with one of your own.

Research, observe and add your discoveries after the question.
Why did you chose this problem
(if it was your choice?

MATERIAL LIST:

Everything should be listed here that will be used. Include the exact quantities.

This is important for others to replicate your experiment in the future. If you needed more or less during the experiment, document this in your procedure/results.

HYPOTHESIS:

DATA TABLE:

Include whatever data table you use. Graphs of results do not go here; they go into analysis if you do them. Make the data table clear and brief.

PROCEDURE:

Use starting words such as obtain instead of get. Do not start each one with First, I went and got...etc. Try using the word, "investigator" or "researcher". Remember another person may have to duplicate your lab, so it must be written in a clear, brief manner that will help prevent error in the lab.
Reproducibility is important for your experiment findings to be valid.

ANALYSIS:

Look over your data, make and include graphs here, determine if your experiment supports or refutes your hypothesis. Don't worry if your hypothesis is not proven. The experiment is as valid as one in which hypothesis is proven be correct.

SUMMARY

This section must contain each of the items listed below. You are now the one speaking, of your personal results. What you did, how you did it, why you chose to do it in that way, what you learned, possible error/flaws of the lab (you must include at least one), how you would change it in the future to enhance, improve or due to changing hypothesis.

NOTES FOR EACH HEADER SECTION

Title: Make your title unique and memorable!
Question/Problem:
Material List:
Procedures: Use starting words such as obtain instead of get. Do not start each one with First, I went and got...etc. Try using the word, "investigator" or "researcher". Remember another person may have to duplicate your lab, so it must be written in a clear, brief manner that will help prevent error in the lab. Reproducibility is important for your experiment findings to be valid.
Data Table: Include whatever data table you use. Graphs of results do not go here; they go into analysis if you do them. Make the data table clear and concise.
Analysis: Look over your data, make and include graphs here, determine if your experiment supports or refutes your hypothesis. Don't worry if your hypothesis is not proven. The experiment is as valid as one in which hypothesis is proven be correct.

You could begin like this:
 In this lab I _____because_____ . I
did this by _____.
I found out/learned that _____.
 Some errors that may have occurred with this lab include
_____. In
 the future I would (change, add, delete)_____to enhance the lab.

OBSERVATION CHALLENGE

OBSERVATION CHALLENGE:

FORENSICS WORD SEARCH

```
E F E P Y X W X J B C K K Y I
H V J C S E T F A X T Q N Z C
A P I A A C W L K N E V O N F
N X T D E R L N I J K E C U V
A Q X P E I T R W I T N E S S
L H S L S N P F Q P W X E C D
Y U H T D R C A U T O P S Y O
S O I F E N J E K S A T G M O
T C Y G O L O C I X O T T E L
S P N P Y R O T A R O B A L B
R I R F H D E O Z P D T C X A
F M N Y K B U N V Z O A R C C
N O I T A G I T S E V N I J E
J X Y V U I B Q O I F D M V T
V C Q S T O E C Q Z C S E R Y
```

ANALYST	AUTOPSY	BALLISTICS
BLOOD	CRIME	DNA
EVIDENCE	FINGERPRINT	FORENSIC
INVESTIGATION	LABORATORY	SUSPECT
TOXICOLOGY	TRACE	WITNESS

Can You

Spot the differenc?

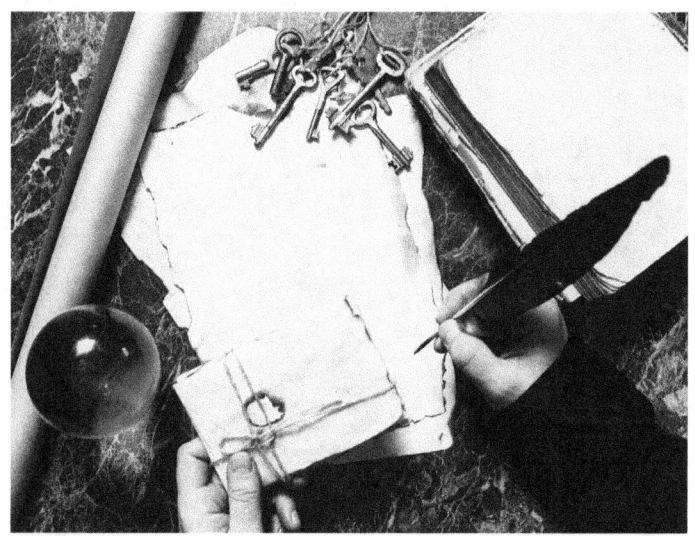

CSI WORD SEARCH

Words can go in any direction.
Words can share letters as they cross over each other.

Look for the words listed below.

```
E  N  Y  E  P  D  Y  Q  T  P  L  L  T  A  A
N  V  S  S  E  N  T  I  W  W  N  I  A  T  U
Z  O  I  B  L  O  O  D  S  K  O  R  N  B  T
N  Z  I  D  K  B  O  K  F  W  R  I  Y  E  O
K  Q  L  T  E  K  N  Z  F  F  R  R  Y  V  P
V  R  U  B  A  N  T  M  O  P  R  O  P  I  S
O  X  T  G  B  G  C  S  R  Y  U  A  B  T  Y
C  A  S  E  E  K  I  E  E  U  S  N  B  C  H
E  N  E  C  S  F  G  T  N  U  R  D  R  E  D
O  U  U  X  E  N  Q  S  S  Z  R  I  J  T  A
Y  Z  H  V  I  U  X  P  I  E  M  F  K  E  S
K  E  L  F  I  A  E  S  C  E  V  Q  H  D  C
F  O  L  Z  W  C  T  Q  Q  Y  L  N  R  O  Y
S  U  T  G  T  E  H  O  P  I  P  L  I  K  N
E  K  U  F  P  C  N  J  E  J  R  J  Q  J  Y
```

EVIDENCE
FORENSIC
CRIME
-SCENE
INVESTIGATION
DNA
FINGERPRINT
BLOOD

LAB
AUTOPSY
SUSPECT
WITNESS
DETECTIVE
SOLVE
-CASE

Mystery of the Labyrinth Maze:

CSI
BINGO

CSI
BINGO

CSI BINGO CALL SHEET

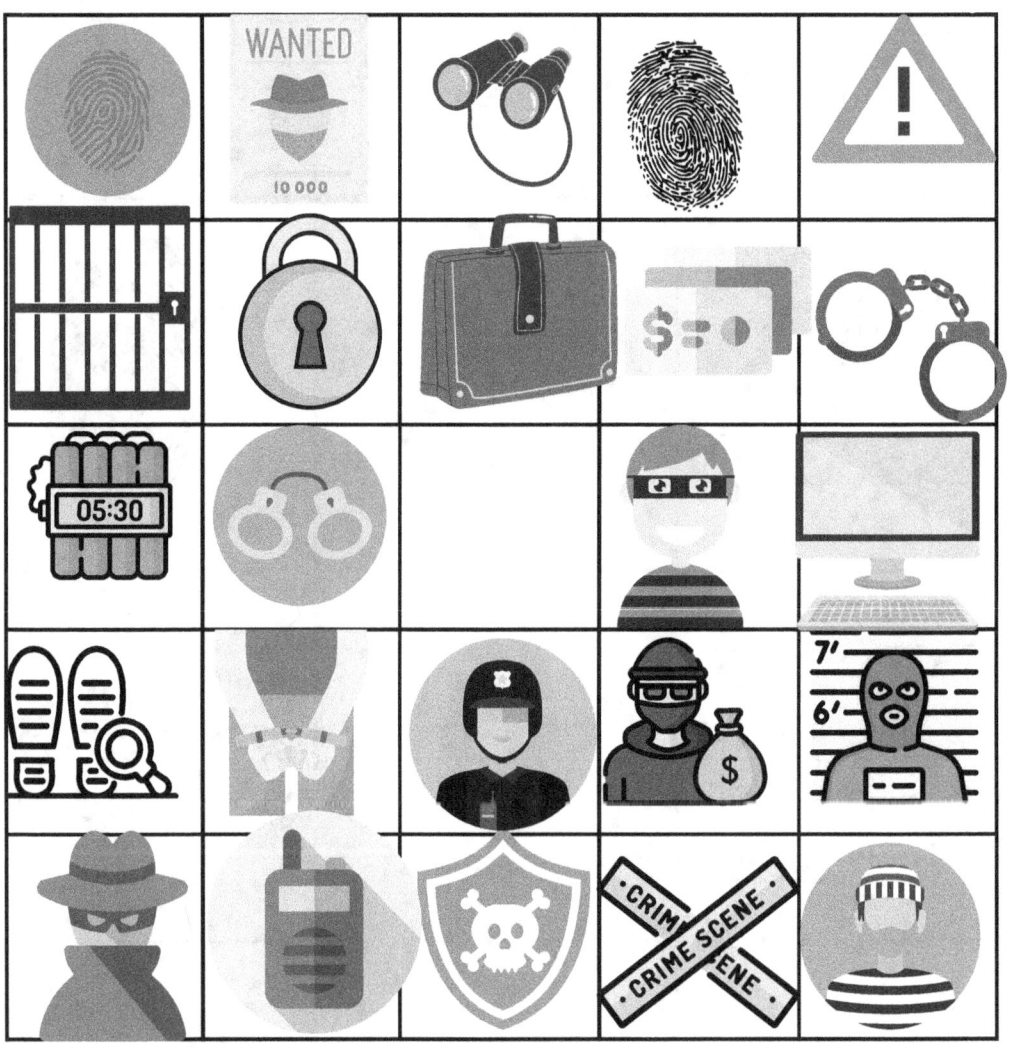

CSI CODE BREAKER I SPY GAME

Code breaking, I Spy activity where participants decipher hidden messages related to King Solomon's wisdom, linking the ancient wisdom to modern investigative techniques.

"I spy with my code-cracking eye, something that begins with..."

1. S: Find a scroll or ancient-looking document symbolizing the connection to historical codes.

2. C: Spot an item that represents a key or code sheet, essential for deciphering hidden messages.

3. W: Identify a wisdom symbol hidden within the game, linking to King Solomon's teachings.

4. I: Locate an item used for invisible ink, adding an element of mystery to the code-breaking.

5. P: Spy a puzzle piece that needs to be assembled to reveal a crucial part of the code.

6. C: Discover a cipher wheel, an essential tool in the art of code-breaking.

7. M: Uncover a magnifying glass to aid in scrutinizing intricate details of the codes.

8. K: Find an object or symbol resembling a keyhole, suggesting the need for a decoding key.

MAZE

CATCH THE BANK ROBBER

CRIME SCENE REPORT

CRIME SCENE

On Inday afternoon at 3:30 pm, The First Bank & Trust was robbed. This image was captured on the bank surveillance camera.

1

[ ]

[ ]

1. Observation
   2. Senses
   3. Memory
4. Equipment
   5. Tools
6. Notebook
   7. Pencil
      8. Pen
2 9. Ruler
10. Measuring tape
   11. Magnifier
12. pH test strips
13. Blood detector
14. Fingerprinting

2 In your own words, describe what happened during the crime.. Include details like:How many criminal? What were the details of the robber? Remember who, what, when & where.

_____

_____

_____

_____

_____

_____

Properties of   MATERIALS

Take a walk around your classroom or playground.
Circle the different materials as you find them.
Then complete the table.

concrete	paper	wood	fabric
glass	metal	plastic	rubber

object:	looks like: 👁	feels like: ✋

object:	looks like: 👁	feels like: ✋

object:	looks like: 👁	feels like: ✋

object:	looks like: 👁	feels like: ✋

Match & Write

Directions: Match the image with the first letter of the word. that describe it Trace the letters accordingly.

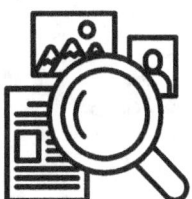

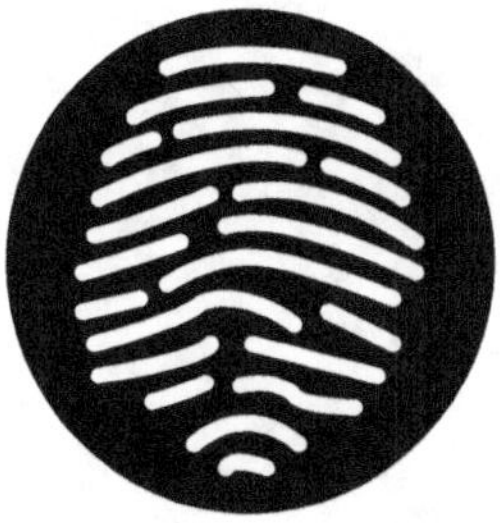

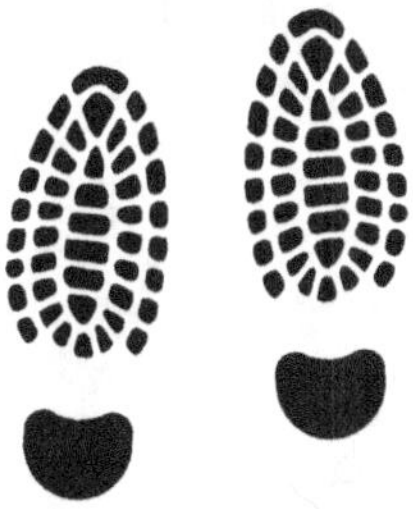

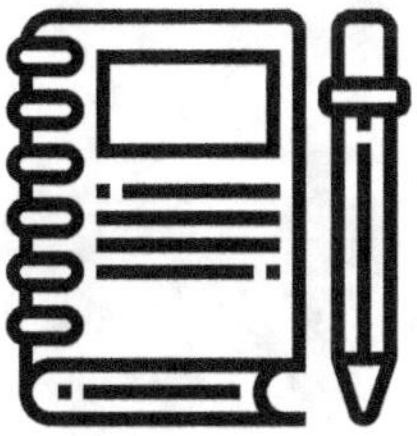

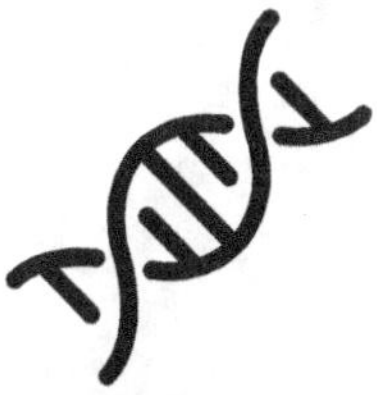

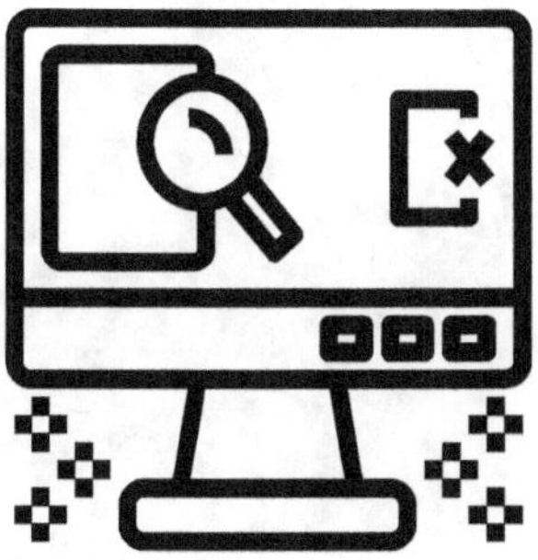

Famous Detectives
in Literature Word Search

```
A  T  E  H  N  T  P  F  L  E  T  C  H  E  R
T  S  L  N  I  N  W  O  R  B  Z  Y  S  Q  M
M  O  P  F  J  E  S  S  I  C  A  H  A  I  A
N  O  R  E  N  M  W  A  W  R  E  P  U  N  R
I  S  A  I  A  I  E  E  S  R  O  E  G  V  P
I  S  M  S  O  U  R  J  L  A  H  T  U  I  L
Z  N  T  R  B  P  D  O  N  I  D  E  S  N  E
H  O  L  M  E  S  C  I  N  V  K  R  T  L  N
E  D  A  P  S  K  S  E  M  L  O  H  E  I  B
W  S  A  W  H  E  R  C  U  L  E  S  P  W  I
I  L  S  T  O  D  F  U  Q  B  Y  U  A  N  A
M  J  G  I  U  L  R  L  J  R  D  C  L  S  L
S  P  Q  P  M  S  F  O  O  O  S  H  N  U  A
E  J  I  N  I  I  Z  E  L  W  A  R  J  A  S
Y  N  S  P  A  D  E  Z  R  N  S  A  S  J  N
```

1.. FAMOUS DETECTIVES IN LITERATURE:
2. HOLMES (AS IN SHERLOCK HOLMES)
3. POIROT (AS IN HERCULE POIROT)
4. MARPLE (AS IN MISS MARPLE)
5. DUPIN (AS IN C. AUGUSTE DUPIN)
6. SPADE (AS IN SAM SPADE)
7. WOLFE (AS IN NERO WOLFE)
8. FLETCHER (AS IN JESSICA FLETCHER)
9. DREW (AS IN NANCY DREW)
10. WIMSEY (AS IN LORD PETER WIMSEY)
11. BROWN (AS IN FATHER BROWN)

ANSWER: CSI MEMORY CHALLENGE

ANSWER - SCIENTIFIC METHOD WORD SEARCH ANSWER

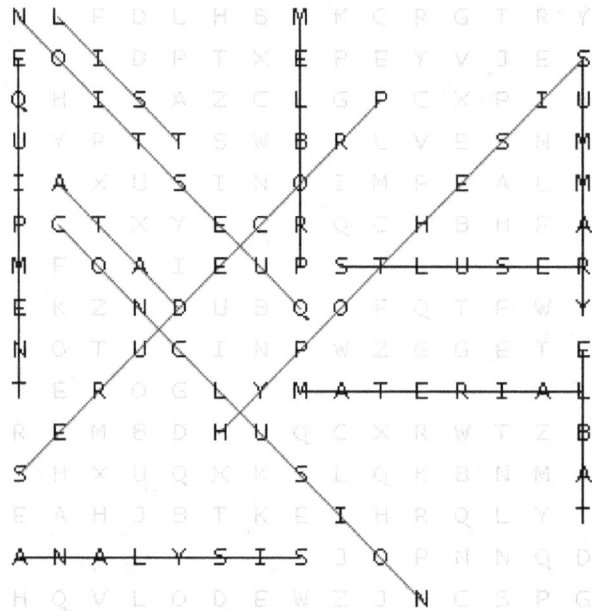

ANSWER - SKILLS & TOOLS OF AN INVESTIGATOR WORD SEARCH

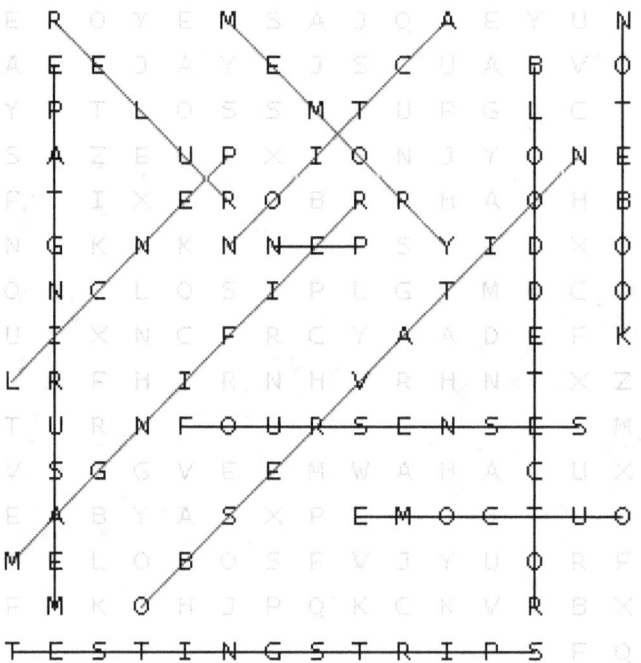

ANSWER - INVESTIGATOR TOOLS WORD SEARCH

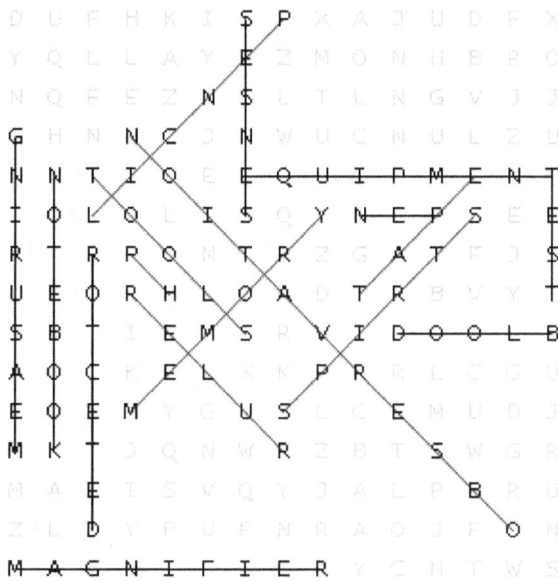

ANSWER - FORENSIC SCIENCE TECHNIQUES

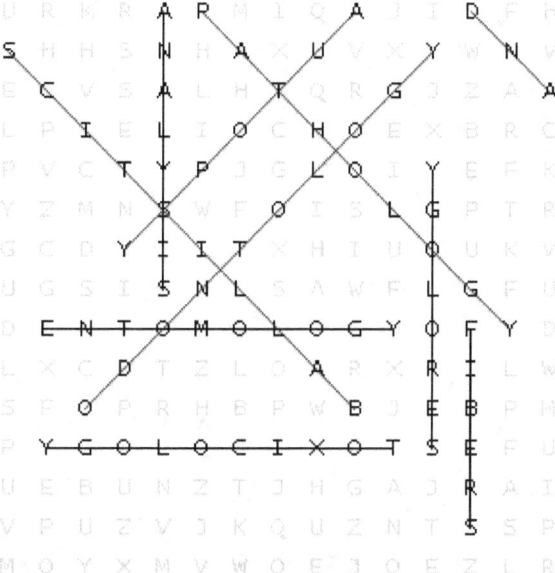

ANSWER - FAMOUS LITERATURE DETECTIVES

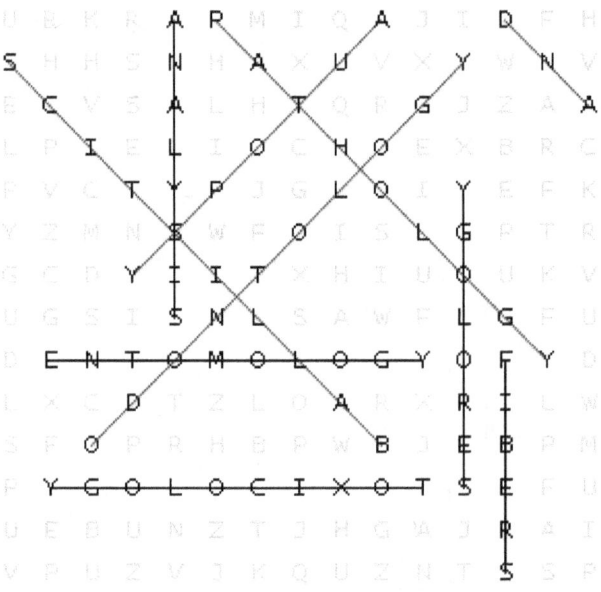

CSI HUNT ANSWER

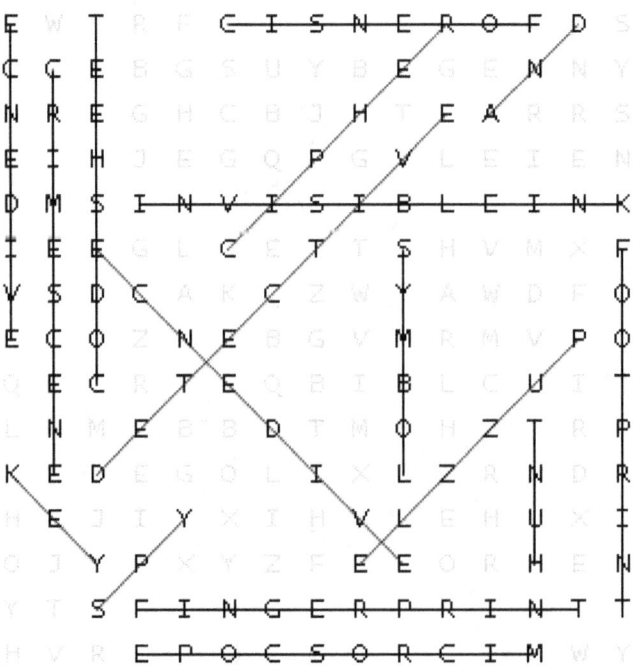

ANSWER - TO CATCH THE BANK ROBBER MAZE

ANSWER - STOP THE CYBER THIEF MAZE

ANSWER - AMAZING DNA

Mystery of the Labyrinth:

Answer: Observation

Challenge What do you see?

1. Pair of Knives
2. Cigarette
3. Ashtray
4. Face Disguise
5. Poision
6. Needle
7. Gun

Answer- CSI Word Match

1. Autopsy (Postmortem examination to discover the cause of death)
2. Ballistics (The study of the effects of being fired on a bullet, cartridge, or gun)
3. Contamination (Unwanted pollution of a crime scene, which can obscure evidence)
4. DNA (Genetic blueprint used for identification in forensics)
5. Entomology (The use of insects to help solve criminal cases)
6. Fibers (Threads or filaments from fabric used as evidence)
7. Fingerprint (Unique pattern used for personal identification)
8. Odontology (Study of teeth structure and dental work for identifying human remains)
9. Pathology (The study of disease and its effects on the body)
10. Toxicology (The study of adverse effects of chemicals on living organisms)

Scavenger Hunt Answers

Answer Wisdom in Nature Scavenger Hunt I Spy List:
- Wise owl figurine or image
- Tree with rings (representing age and wisdom)
- Beehive (symbolizing organization and order)
- Ant trail (representing diligence)
- Spider web (symbolizing intricacy and design)
- Rock with interesting patterns (connecting to geological wisdom)
- Water droplet (symbolizing purity and clarity)
- Plant with unique growth patterns (representing adaptation and growth

Answer CSI Code Breaker:
- CSI Code Breaker I Spy List:
- ·Scroll (representing ancient text)
- ·Code sheet
- ·Cipher wheel
- ·Invisible ink
- ·Key or keyhole
- ·Wisdom symbol
- ·Puzzle piece
- ·Magnifying glass

Forensic Evidence Hunt I Spy List:
·Magnifying glass
·Fingerprint
·Crime scene tape
·Microscope
·Evidence bag
·Detective badge
·Footprint
·DNA model

www.ingramcontent.com/pod-product-compliance
Lightning Source LLC
Chambersburg PA
CBHW080449290526

45791CB00008BA/2650